SILKWORMS

L. PATRICIA KITE

Illustrations by
Karen Jeanne Kite

Cover photo by Ron West

LPK Science
Newark, California
1997

ISBN: 1-57502-542-6

Library of Congress Catalogue Card Number: 97-94030

Published by

LPK Science
5318 Stirling
Newark, CA 94560

Printed in the United States by
Morris Publishing.
3212 East Highway 30
Kearney, NE 68847

TABLE OF CONTENTS

This book is dedicated to Joyce Peters, owner of Giant Steps book store in Fremont, who urged me to write a new book about Silkworms, and to so many others who encouraged persistence in getting it published: Peggy Seaman, Zoe Siegel, Jason Carter, Chris Souza, Margery Ulrich, Clarice Weisbach, Robin Michel, Jane Burns, Linda Vernon plus teachers everywhere.

Chapter One

EMPRESS HSI LING-SHI AND THE SILKWORM

There are several legends on the beginning of silkworm use for human clothing. All legends begin in the Chinese royal palace, about 4,635 years ago.

An often-told story has the Emperor Hwang-Te complaining that something was eating leaves off the beautiful mulberry trees in his palace gardens. He instructed Empress Hsi Ling-Shi (also known as Xi Ling Shi) to find out what creature was doing this mischief.

The Empress watched carefully. It didn't take long for her to see that white caterpillars were eating leaves day and

night. Then Hsi Ling-Shi noticed a hard white cocoon attached to a leaf. She removed and studied it.

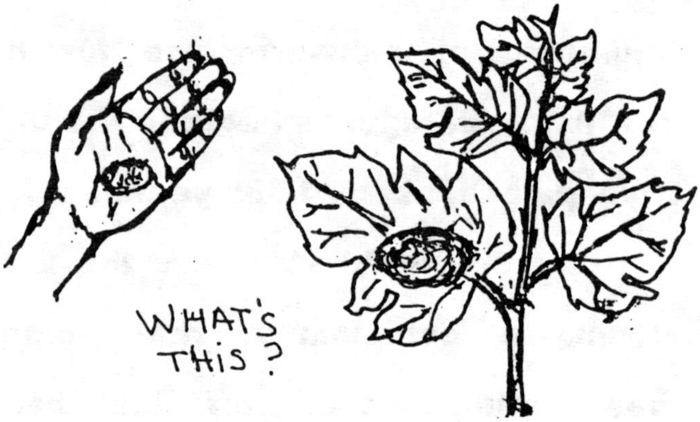

She found other cocoons that the caterpillars had made. The one day she accidentally dropped a cocoon into a cup of hot tea.

Taking the cocoon out of the tea, the Empress saw the cocoon had softened, and a strand of thread had slightly separated.

Playing with the separated strand, she noticed it unwound as one continuous strand. Could this lovely silken thread, called "sse" in Chinese, be used for sewing? It was much stronger than any thread they already had.

Hsi Ling-Shi asked the Emperor for her own special mulberry tree grove. She wanted to raise thousands of silk-producing caterpillars. Of course, the Empress got what she wanted. She got many helpers too.

At first Hsi Ling-Shi and her helpers experimented with the silk threads in secret, studying ways to join the threads together to make small fabric pieces. After much trial and error, she was able to make a beautiful silk scarf, presenting this as a special gift to her husband.

Soon the entire Chinese royal household wore silken robes embroidered with dyed silk threads. The Emperor quickly realized this lightweight, strong material could be sold outside of China. This would help the royal budget a lot.

The Emperor also realized that if other countries knew how to raise silkworms, or make silk, they wouldn't buy it from China. So he issued a law: anyone who told the secret would have his head chopped off.

THE SILK ROAD BEGINS

Silk sold very well to royalty and rich people everywhere. Some even added threads of silver and gold to the beautiful material. The now-famous Silk Road began. It wasn't really one road, but many roads crisscrossing each other. Other valuable items, besides silk, were also carried along the Silk Road.

Traders from ancient Persia bought the brightly dyed silk from China, and took it on caravan to other countries. One horse could carry 600 pounds of silk. One camel could carry 1000 pounds of silk. There were many horses and many camels, and a hard-working trader could become very wealthy.

But it was a very dangerous journey. There were mountains and very hot deserts to be crossed. There were many robbers lying in wait to ambush these traders. But a trader completing the long journey would earn twice what he paid for the silk in China. Many took that risk.

STEALING THE SECRET OF SILK

For 2000 years, the secret of silk remained a secret. Spies of all kinds, some paid by other kingdoms, tried all kinds of tricks to discover the truth. Nobody succeeded.

In the meantime, there were all sorts of rumors. Many thought silk came from the fruit of the tropical silk-cotton tree, Ceiba pentandra. This fruit looks like long leathery capsules. Each capsule is filled with seeds surrounded by silky white fibers. These fibers, often called "kapok," could be used as mattress fillings. But they were certainly not clothing silk.

While a few people mentioned that silk might have been spun by a spider or beetle, most had much stranger thoughts. Perhaps silk came from the hair of sea sheep?

Or perhaps silk was created by worms living in rotten calf meat.

Or perhaps silk was a fleece growing on a tree?

In the meantime, China was making a lot of money selling silk to Persian traders, who were making a lot of money selling it elsewhere. Other countries were very jealous.

Then Roman Emperor Justinian (483-565 A.D.) got an idea! Caravans traveled the Silk Road, but so did monks and pilgrims carrying religious messages. In 552 A.D., Emperor Justinian sent two monks to China. These monks looked like all the others, plainly dressed and carrying bamboo sticks as a walking aid.

Except these monks had a different message. Their instructions were to find out about silk making. Somehow they managed to hide in the tree branches overlooking the royal gardens.

When these monks left China, they looked as plainly dressed as when they entered. They also carried their bamboo walking sticks. Except in the bamboo's hollow center were silkmoth eggs and mulberry tree seeds. The monks had also observed and memorized the technique of silkworm breeding and silk reeling.

The monks went directly to Emperor Justinian's court in Constantinople, now called Istanbul, Turkey. They told all they knew. The secret of silk was no more.

BITS AND PIECES

The Summer Palace Museum in Hanzhou, China, contains embroidered silk pieces estimated to be 4,500 years old.

Recent archeological findings question the 2460 B.C. legend of Empress Hsi Ling-Shi. Some evidence shows silk was used at a much earlier date. However Empress Hsi Ling-Shi is still remembered as the "Goddess of the Silkworms."

When mulberry leaves appear in China today, there may be local festivals recalling the wonderful discovery of silk.

Can you have a classroom festival about silkworm discovery? Everybody could bring to class something which includes silk, even just a picture of something made of silk.

kimono scarf necktie dress fishing line

stockings surgical sutures sari

computer ribbons sewing thread

You might do a play about the discovery of silk and its theft. If the Chinese New Year comes about when mulberry leaves are budding, taking the silkworm eggs from the refrigerator is another way to celebrate.

Chapter Two

SILKWORMS COME TO THE NEW AMERICAN COLONIES

From England to America

King James I tried to start a silk industry in England when he took over the throne in 1603. He enjoyed silk fabrics, as did the Queen and palace staff. But there was an even greater reason for starting a silk industry. Many refugees were entering England at this time, seeking religious freedom. James I figured that if they had jobs, they wouldn't be such a burden to the community. Silkworm farming and silk production would provide these jobs. And the silk could be sold, bringing in needed monies to England.

It started out as strictly a business idea, but James I soon thoroughly enjoyed watching his Bombyx mori silkworms grow. He ordered royal gardeners to plant lots of black mulberry trees on the palace grounds. And when James I had to be away from the garden, he employed special attendants to follow him around carrying a container of silkworms.

The king also offered mulberry seeds and tree seedlings at a special price to palace employees.

James I wanted English colonies in the New World to start a silkworm industry. The first would be Jamestown, Virginia, the settlement colony named after him.

The king ordered several silkworm and mulberry shipments sent by sea to the colony. The first few ships didn't make it. But later ones did. By 1619, if a landowner hadn't planted mulberry trees, there was a possible penalty from the Crown. To stay in good graces, large landowners planted as many as 70,000 black mulberry trees.

But there were continuing problems. Virginia spring and summer weather tends toward the hot and humid. The silkworms didn't thrive in this climate. But tobacco did. Virginia colonists could market the tobacco themselves, while silk thread had to be sent to England to be spun into fabric.

So the Virginia colonists did what was best for them. After a while, only thriving black mulberry trees remained as a souvenir of the King's best efforts.

Other colonies also tried silkworm farming at Britain's demand. But rice and indigo were more profitable in South Carolina. And while Georgia tried silkworms a while longer than the rest,

the weather, and cotton-growing profits, eventually won out.

However in 1725 the silk industry sprang up in the North-East states. Benjamin Franklin, who did business in Philadelphia, was a chief advocate. Extra monies were offered to persons raising silkworms. White mulberry trees were imported from China. Millions were planted. There was almost a "silkworm mania."

Rhode Island, Connecticut, Maine and New Hampshire joined in. The first American silk mill was built in Mansfield, Massachusetts in 1810. Some silk was exported to England. Skilled workers were brought in from Europe. But by 1880, for various reasons, production dropped from 310,000 pounds to 5,000 pounds.

While sericulture, or the breeding & rearing of silkworms to make raw silk, seemed like a good idea from time to time in other places, it never again took strong hold in the United States.

However the Shakers, a religious community living in Kentucky, raised silkworms commercially until 1922.

Chapter Three

RAISING SILK WORMS

Equipment

Mulberry silkworm (Bombyx mori) eggs

A shoebox or other sturdy growing container. An old fishtank is fine.

Mulberry leaves

OBTAINING SILKWORM EGGS

The best time to begin raising silkworms is in March or April. Some teachers do this during Chinese New Year, depending on mulberry leaf availability. However you may order silkworm eggs in

January or Febuary, or even earlier. Silkworm eggs are sometimes available from local grammar schools. The best person to ask is the science teacher.

You may order silkworm eggs from biological supply companies. Check your telephone book or ask a science teacher at any grade level. Prices vary. But once you get an initial egg batch, you will seldom have to order silkworm eggs again. Silk moths produce a lot of eggs. You will have plenty and some to share.

Before you begin raising silkworms, you must have access to mulberry leaves. Silkworms only feed on mulberry leaves, although some people claim they eat lettuce. But they may starve and will not spin without a mulberry leaf diet.

Look around your neighborhood for a mulberry tree (see chapter 5). Have an adult ask the owner of the mulberry tree if you can cut leaves for your silkworms. Don't do this without permission. Silkworms are small, but they eat a lot of leaves as they grow.

A caution: When you order eggs from a supply house, if the weather is warm, sometimes silkworms start to emerge while in the mail. If you don't have mulberry leaves ready, this could be a problem.

But perhaps your mulberry tree doesn't have leaves yet. So be prepared. When you know your silkworm eggs are arriving, cut some mulberry tree branches that have large buds growing on them.

Put these branches in a container of water. Keep this indoors. House warmth will force the buds to leaf out more quickly than in the colder outdoors.

When Silkworm Eggs Arrive

When the silkworm eggs arrive from a supply house, they will be in some sort of a packet. Now you have a decision to make. Look at your mulberry tree. Does it have leaves, even little ones, or leaf buds? If so, you can place your new silkworm eggs in a shoebox or other rearing container.

But don't put mulberry leaves in with the eggs. Cut leaves give off carbon dioxide. This is harmful to the unhatched eggs. Wait two or three weeks until silkworms come out of the eggs before adding mulberry leaves to the container. Put the leaves in as soon as you see the baby silkworms.

What if your mulberry tree has no leaves or leaf buds? Then the silkworm eggs must go in the refrigerator to keep them from hatching until leaves appear.

You may want to place the silkworm eggs in a sandwich bag or small paper bag when placing them in the refrigerator. Not everybody in the family likes to look at silkworm eggs when reaching for a peach. Label the bag, or it might get accidentally thrown out.

Do <u>not</u> place silkworm eggs in the freezer. This will kill the eggs.

A note: Should your eggs arrive from a friend or teacher, they may be on a cocoon or piece of cardboard. Do not remove the eggs, as they will break open.

THE SILKWORM CONTAINER

What size growing container should you use? That depends on how many silkworms you are raising. A shoebox holds about six full-size silkworms. Usually you raise many more. But you can find larger containers later.

Where should you place your container? A bathroom counter or school display table is a good place. This lets you easily watch your silkworms grow.

The container site should not be too hot or too cold. But it should get some light. Silkworms prefer about eight hours of light a day. This can be natural light lamp from a light. But don't put the

growing container too close to a sunny window, lamp bulb, or other heat source. The silkworms will bake.

If a garage gets light during the day, or artificial light is available, you can place silkworms here. But this is only true if no cars go in and out, the hot water heater is not in the garage, and no chemicals are stored here. Car exhaust, natural gas fumes, and any insect sprays will destroy your silkworm project.

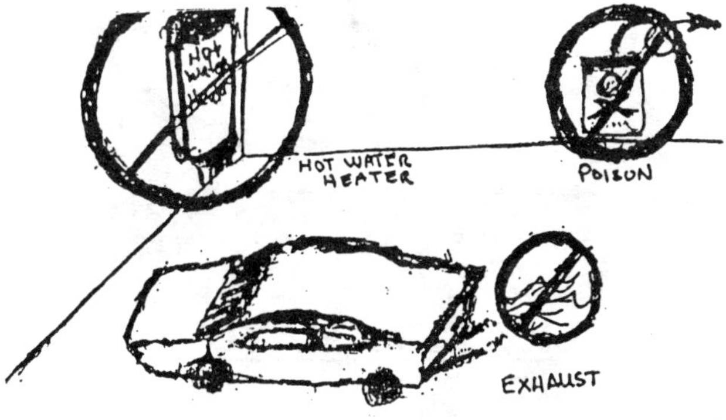

Silkworm protection is necessary inside inside your home too. The site you select must be watched for ants. Household ants will attack and destroy helpless silkworms and silk moths.

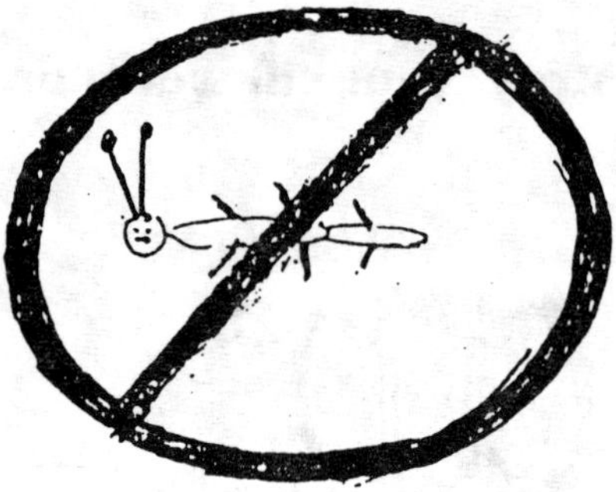

SILKWORM HATCHING

Silkworms will hatch from the eggs about 14-21 days after removal from the refrigerator. Watch the eggs change color in the meantime. They keep getting darker, especially around the edges.

Once hatching starts, it can go on for up to a 14 day period. Each egg has its own readiness time.

You must look very carefully to see newly hatched silkworms. They are about 1/8-inch long, and look like teensy black threads. But you can see them easily with an inexpensive magnifying glass.

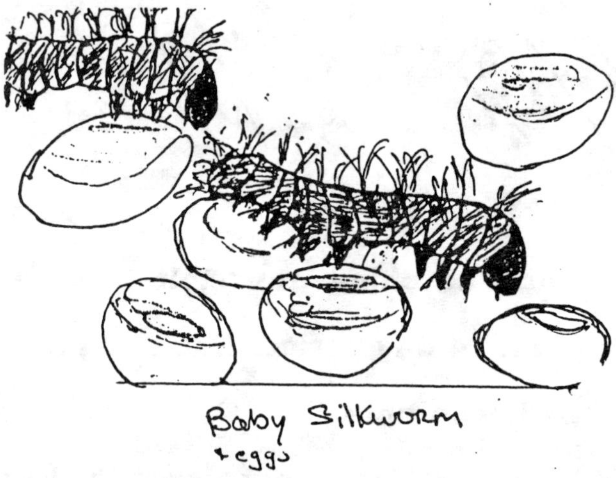

Remember: Silkworms need to eat as soon as they come out of the egg. Place some shredded mulberry leaves in the container.

SILKWORMS GROW BIGGER

In about a week the silkworms become white or greenish white. If you listen, you can hear the crunch-crunch as they eat.

Silkworms will eat day and night if you give them enough food.

In order to keep growing, silkworms <u>molt</u> or shed their skins (see chapter 4) When a silkworm is shedding skin, it holds on to a leaf, keeps its head up and does not move about for a short while. A silkworm can be injured if you pick it up when it is shedding skin. But you can watch it with a magnifying glass.

Some of your silkworms will grow faster than others. But eventually all become about 10,000 times larger than they were at birth. A silkworm grows to almost 3 inches long and a 1/2 inch wide. This is about the size of an adult's finger.

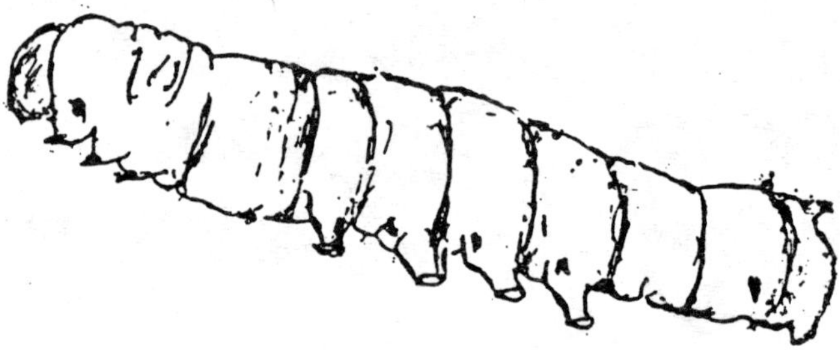

Most silkworms are a creamy white color. But very rarely you may see different colors. If you see a different color, you are very lucky. Silkworms may be oyster white with black bands; a pale blue-white, or brown. Of greatest rarity is the yellow caterpillar. But it does not make a cocoon. It is believed the brown silkworm is closest to the orginal extinct wild ancestor.

*****

Silkworms soon seem like pets to many people. They don't seem to mind if you touch their bodies gently or stroke their head lightly. They are very soft and smooth to the touch. The only time being touched disturbs them is when they are molting.

CLEANING THE CONTAINER

Because silkworms eat so much, they give off a lot of waste material. This blackish waste material, or frass, usually falls to the container bottom. It can make quite a pile, especially when mixed with leaf bits the silk silkworms haven't eaten.

You will have to judge for yourself whether to clean the container. Some silkworm owners just leave the waste there until the silkworms have gone into the cocoon stage. But if the material starts getting a grey covering, it is getting moldy. For silkworm safety, you will want a clean container.

Some silkworm owners routinely gently move the silkworms to a clean container from time to time. Others move the silkworms temporarily, empty the container of waste material, and replace the silkworms.

The best way to move the silkworms is to move the leaves or twigs they are sitting on. That way they are not being disturbed during a molt, or skin shedding period.

SPINNING THE COCOON

When silkworms are about four weeks old, you will see a few of the largest silkworms shaking their heads from side to side. You notice they have stopped eating. You also may see what looks like green waste material instead of the usual black. Any food recently eaten has not been digested, so it has a green leaf color.

Silkworms may also do what is called "emptying their gut." This is throwing up any remaining food within the body.

The silkworm is getting ready to make its cocoon. At this time, silkworms may crawl to odd places on the container, such as the container sides, corners and tops. On occasion, they may even crawl out of the container and start a cocoon on a nearby countertop or the container outside.

To avoid silkworm escape, you may want to place a lightweight covering over your container. Make certain it has airholes. A silkworm may create its cocoon on the covering undersides, so always lift the covering carefully.

SILK THREADS

At first, you will see some thin white threads near the silkworm. These threads are anchors for the cocoon. Then the silkworm makes the cocoon around itself. It starts with the cocoon top layer.

Each silkworm moves its head in a figure-8 pattern as it places silk to make the cocoon. It also gives off a sticky fluid, sericin, which helps the threads on the cocoon stick together. (see chapter 4).

Soon the silkworm is working from inside the cocoon.

THE COCOON

It takes about 2 to 4 days of constant work before the cocoon is completed. Never disturb a silkworm spinning its cocoon. A cocoon is created from a single silk thread. This is about a mile long. A mile is about 10 city blocks.

If the silkworm is disturbed, the thread may break. The silkworm may then try to start another cocoon. But this one can only be partially completed because the silkworm no longer has enough silk. Silkworms seldom form into moths within a partial cocoon.

Inside the cocoon, the silkworm begins its change into a pupa (see chapter 4).

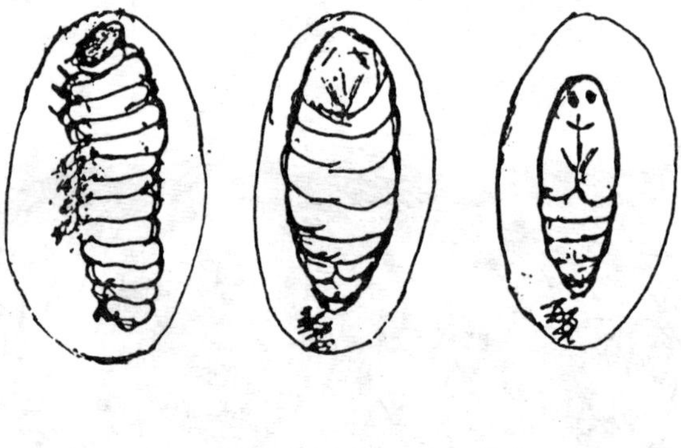

While silkworm cocoons are usually white, they can be yellow, brown or orange. The various colored cocoons make a good science project (see chapter 8).

All your silkworms won't spin at the same time. But within a week, most will have completed their cocoons. Some silkworms fail to grow or don't spin a cocoon. You can continue feeding them, but they will eventually stop eating, completely stop moving, and die.

If you like to examine a dead silkworm, an adult can help you cut it open to see what's inside. (see chapter 4). Work on newspaper and be careful using any type of scissors or cutting tool.

After all the silkworms have spun their cocoons, the container will be very quiet. Within the cocoons, each silkworm is changing into a pupa. Now you can remove the cocoons and give the container a thorough cleaning.

**There is a scientific name for the whole silkworm to silkmoth changing process. It is metamorphosis. The original Latin base for the word is meta, meaning "change."

THE SILKMOTH COMES OUT

When the first silkmoth comes out, it usually sits on top of its cocoon. It is waiting for a silkmoth of the opposite sex to come out of a different cocoon.

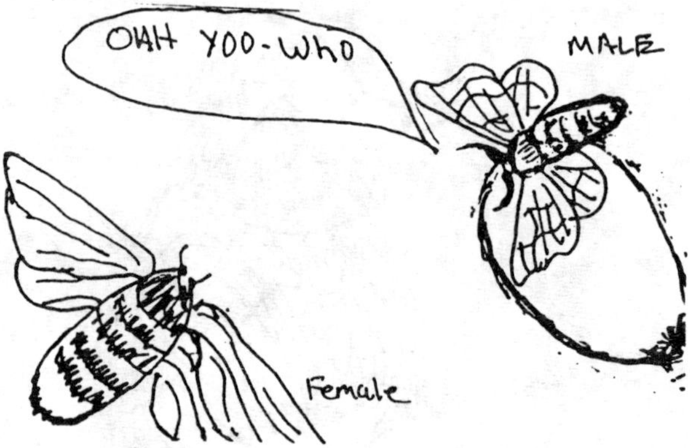

At first, it is not easy to tell male silkmoths from female silkmoths. But when several are present, you can usually see that the female silkmoth has a rounder, larger body. The males are slimmer and their back body part tends to be more pointed. The males also have more feathery-appearing antennae.

Male silkmoths flap their wings almost constantly when they are attracted to a female. This makes a clear whirring sound. When you have a lot of silkmoths, you will want to place the container away from where anybody sleeps. That whirring keeps some people awake.

Many books say silkmoths can't fly. However some get about better than others. Occasionally a male silkmoth gets so excited it hops/flys out of the

container. It may land on the floor, or even manage to get on a nearby wall.

Cats like to catch silkmoths and play with them. This is not good for a silkmoth's health. Keep a cover with airholes over your silkmoth container. If you have a cat, put a book on top of the cover so the cat can't knock the cover off. (But don't cover the airholes)

Do not worry about feeding silkmoths. They do not eat or drink.

SILKMOTH EGG LAYING

Shortly after leaving the cocoon, female silkmoths begin laying eggs. They place these whitish pinhead-size eggs on their cocoons, container sides, and/or container bottom. The eggs have a sticky covering, and stay in place very well.

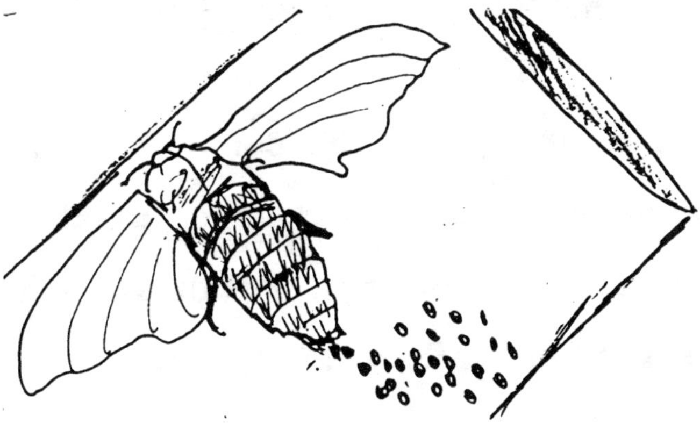

If you place empty toilet paper rolls in the container, the female will lay eggs here too. Usually it is easiest to cut the toilet paper roll in thirds or in half. If the female lays eggs here instead of on the cage sides, the eggs are easier to transfer.

Female silkmoths can lay about 300 eggs each.

STORING SILKWORM EGGS

Silkmoths live about two weeks. When the silkmoths are dead, you will have many eggs in the container. Place the egg-covered cocoons or cardboard rolls in sandwich bags. Put these bags in the bottom portion of the refrigerator.

Silkworms need cold storage before they will hatch next year.

Do not place silkworm eggs in the freezer. If you cannot place silkworm eggs in the refrigerator, you may try keeping them in a very cool place. But if this silkworm egg storage space warms up, the silkworms may hatch too early in the season. If mulberry leaves aren't ready, these early silkworms will starve.

Many students want to know if they can raise more than one generation of silkworms per year. In China, Japan and other silk-producing countries, silkworms are raised all year. Special rooms, lights, heat and food-producing technology are necessary for year-round silkworm raising.

While this special equipment is not available for most homes or classrooms, you can still raise more than one silkworm group while mulberry trees still have leaves. Just don't take all of your silkworm eggs from the refrigerator at the same time. Wait a week, or even two weeks, between batches. That way you will have silkworms for a longer period.

Chapter Four

SILKWORM ANATOMY

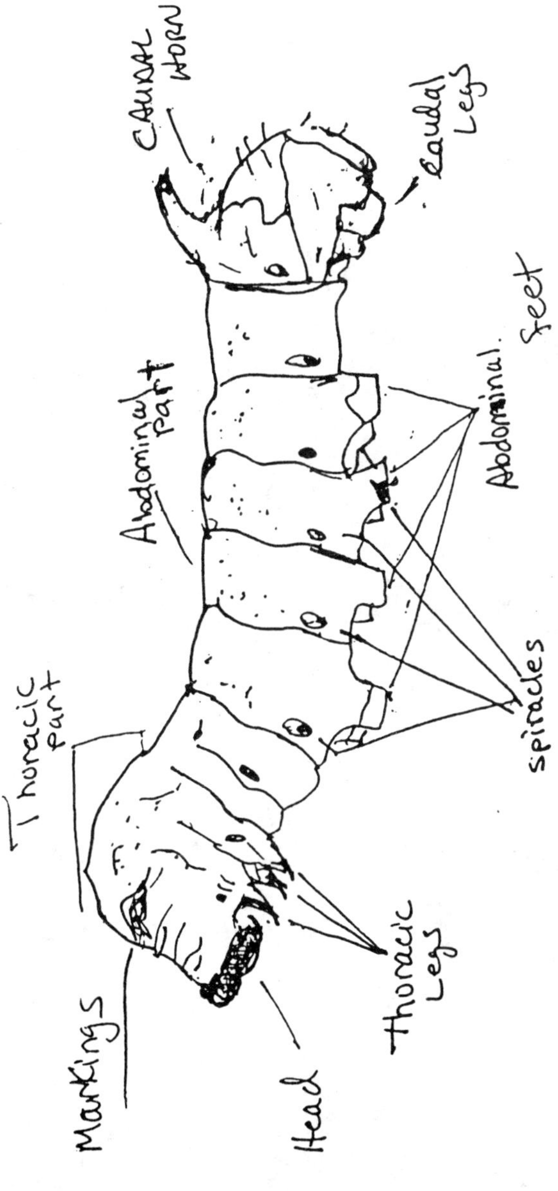

Vision: each silkworm has 12 eyeparts or simple eyes. These eyes see only the difference between light and dark.

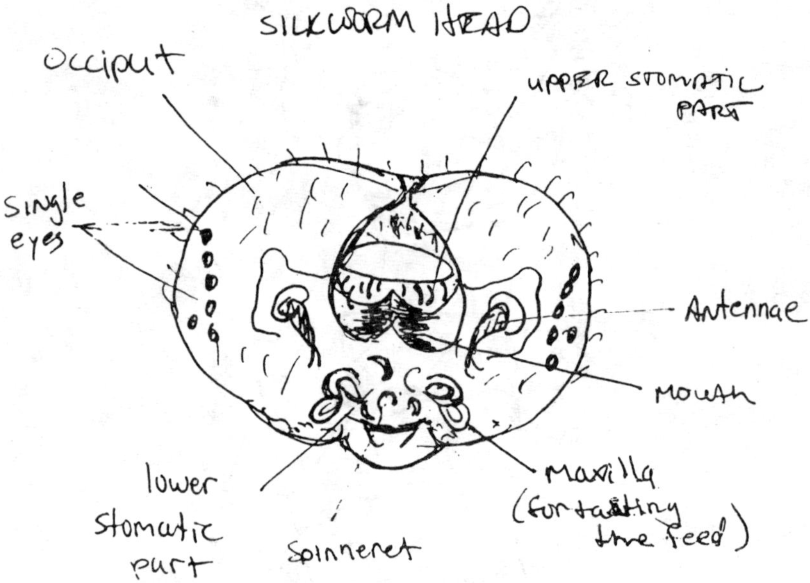

Feeding: Food enters the silkworm's mouth and goes through its digestive tract. Silkworms have a very long gut. Here chewed mulberry leaves mix with body chemicals. The leaves dissolve into smaller and smaller pieces. When the leaves are in liquid form, they pass through the gut and into silkworm body cells to provide nutrients.

Some food material cannot be used. This passes into Malpighian tubes. From here, dark waste matter or frass passes out the silkworm's body.

Breathing: A silkworm breathes in air through its spiracles. These are 9 tiny brown holes or air tubes on each body side. You can see these if you look closely. When a silkworm crawls, it helps bring air into the spiracles. This air enters

the silkworm's <u>tracheae</u>, or internal breathing tubes. From here, air travels to the rest of its body.

<u>Other internal organs:</u> A silkworm has a multichambered heart; a nerve cord; nerve cell ganglia or groupings; and silk glands.

<u>Silk glands:</u> A silkworm's two silk glands are long clear-colored soft tubes. Due to centuries of selective breeding, silk glands are now larger than they were in the wild silkworm. Each gland is now about 12 to 15 inches long. In order to fit, these long narrow glands are curled up within the silkworm's small body.

SILK GLAND

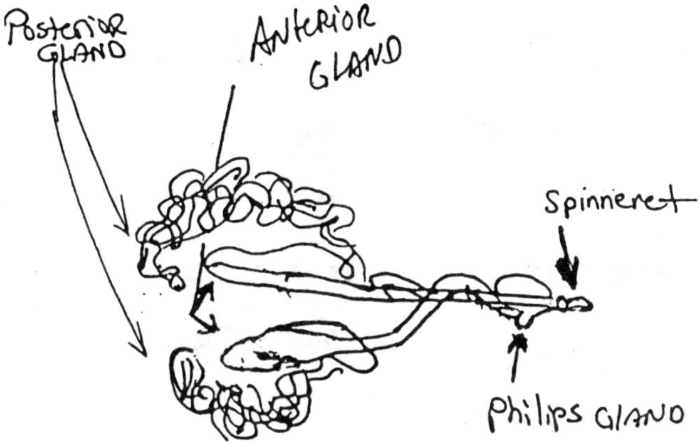

Within each silk gland is silk. But this silk is a thick sticky liquid.

<u>FROM LIQUID SILK TO SILK THREAD</u>

Liquid silk from the two silk glands passes from the silkworm's body through a small tube directly under the silkworm's mouth. This tube is called the <u>spinneret.</u>

As soon as the two liquid threads, one from each silk gland, emerge from the spinneret, they come in contact with air. Each liquid thread now quickly hardens into fine silk thread.

At the same time this happens, the silkworm gives off a sticky material called sericin. The sericin joins the two silk strands together, so they look like one. This silk thread forms the cocoon.

Movement: Each silkworm has six short "true" legs, three on each side of its thoracic segments. These true legs each have a claw. The claw helps the silkworm hold on to leaves or twigs. The three pairs of true legs become the three pair of legs seen on the adult moth.

A silkworm also has 4 pairs of prolegs or "false" legs on its abdomen, or mid-body. These fleshy legs are really outcroppings of the body wall. Sometimes they are called "abdominal feet." At the end of each false leg is a curved hook.

There is also a pair of claspers at the silkworm's rear. The caterpillar uses these to help hold onto leaves and other items.

Growth: **A fully grown silkworm is about 10,000 times larger than a newly hatched silkworm. A silkworm comes out of an egg the size of this dot (.) This silkworm is black and just 1/8-inch long.**

Under a microscope, or a good magnifying glass, you will see the new silkworm has a black head and hairy body.

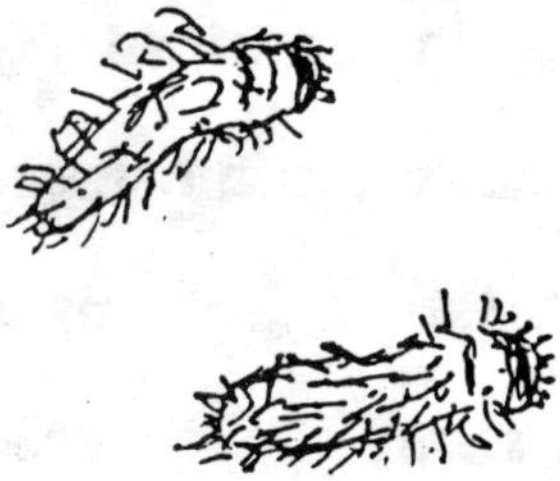

This is scientifically called the first larval or caterpillar stage. But silkworm is the usual word used for Bombyx Mori young.

As a silkworm grows, its skin stays soft, but it doesn't stretch well. So the skin becomes too tight. To continue getting larger, the silkworm will shed its skin, or molt. A new skin is ready underneath the old one.

A silkworm's first molt comes about four days after hatching. This is hard to see, as the silkworm is so small. First the silkworm gives off a bit of liquid silk from its silk glands. It attaches this silk to a leaf. Then it holds onto the silk strands with its claspers, and pulls out of the old skin.

This process is repeated four different times. The second molt usually occurs after another four days, the third after five days, and the fourth molt after six days. Molting is easier to see when the silkworm is larger. What happens to the old skin? The silkworm usually eats it. Then the silkworm goes right back to eating mulberry leaves.

FROM SILKWORM TO PUPA

The time spent living in each skin before it is shed is called an instar. A silkworm has five instars. But the last instar is different than the rest. The silkworm has gotten as large as it can. It stays in the fifth instar about eight days, before it begins spinning a cocoon.

Once inside the finished cocoon, it sheds its skin for the final time. Now the silkworm begins changing form.

First it changes into a yellowish-colored pupa. This takes about 10 days. It gets shorter and shorter. When it is about 2/3-inch long, it begins changing into a silkmoth.

FROM PUPA TO SILKMOTH

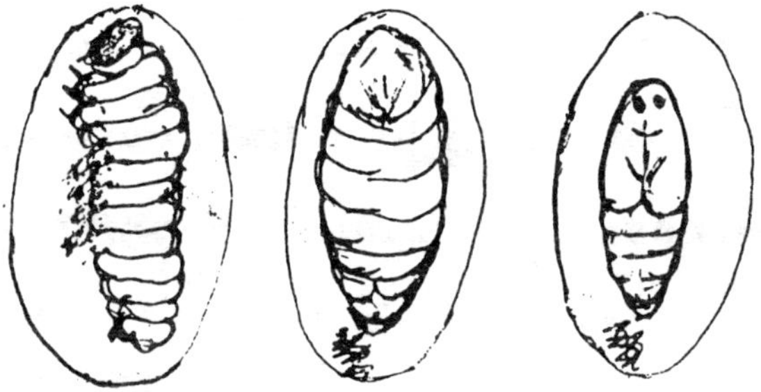

Slowly the pupa becomes dark brown. Its outside hardens. Within the pupa covering, a silkworm's crawling legs disappear. They are replaced by the adult moth's four legs. The 12 eyeparts change into two large black compound eyes. Wings and antennae develop.

During this time, internal organs called ovaries began developing within female silkmoth pupae. The process of egg production begins in these ovaries.

About two weeks from the time each cocoon was completed, the moth within is ready to emerge. This is the end of the fifth or final instar.

THE SILKMOTH COMES OUT

Each silkmoth now gives off a special liquid which softens one end of the cocoon. The area around the hole often turns brown from the action of this dissolving fluid. The silkmoth pushes its way out. You may see this happen. It most often happens early in the morning.

At first, the silkmoth sits on top of its cocoon. Its body is crumpled from being inside the small cocoon. But the silkmoth pumps body fluids into its wings until they straighten out.

Each silkmoth has two feathery-appearing antennae or feelers sticking out from the top of its head. These are sensory organs. They act to sense odors and noise, somewhat like our nose and ears.

The antennae of male silkmoths have more branchlets and therefore appear more feathery. In the wild, these extra branchlets helped the male better sense the female, finding her even if she was some distance away.

Mating activity begins almost immediately after silkmoths come out of the cocoon. Each female gives off a chemical odor, called a pheromone, from special scent glands at the end of her abdomen. These very tiny glands are sometimes visible. The pheromone helps the males locate the females by smell.

Mating takes place abdomen to abdomen, although it looks like rear to rear. Each male's wings flutter strongly during this time, and also whenever a female silkmoth is sensed nearby.

Scales which cover a moth's wings fall off from this wing fluttering, becoming part of the litter on the container floor. Toward the close of the mating process, you will see males with bare spots on their wings because so many of their wing scales have fallen off.

"I HAVE NAKED WINGS"

Eggs move from the female silkmoth's ovaries into special tubes called <u>oviducts.</u> Here they are fertilized by <u>sperm</u>, or reproductive cells, that come from the male silkmoth.

The fertilized eggs exit the female's body through a special opening at the end of her abdomen. These flat eggs are pin-head size. Initially they are a dirty-white color. But all fertilized eggs darken gradually. Any eggs that were not fertilized will stay yellow.

CHAPTER FIVE

ALL ABOUT MULBERRY TREES

There are several different types of mulberry trees. Silkworms prefer the White Mulberry, but you may not have a mulberry tree choice where you live. So you should be able to recognize any type of mulberry tree, since silkworms generally eat nothing else but mulberry leaves.

WHITE MULBERRY

The white mulberry, Morus Alba, is also called "silkworm mulberry." It is native to China. The white mulberry was

brought to North America by the British in their attempts to set up a thriving silk industry (see ch. 2).

White mulberries grow from 20 to 60 feet high. The bark is yellow-brown. White mulberry buds are red-brown. Leaves are 3 to 5 inches long, and smooth on both sides, with coarse jagged edges. All mulberries are deciduous trees, meaning their leaves fall off in winter and reappear in spring.

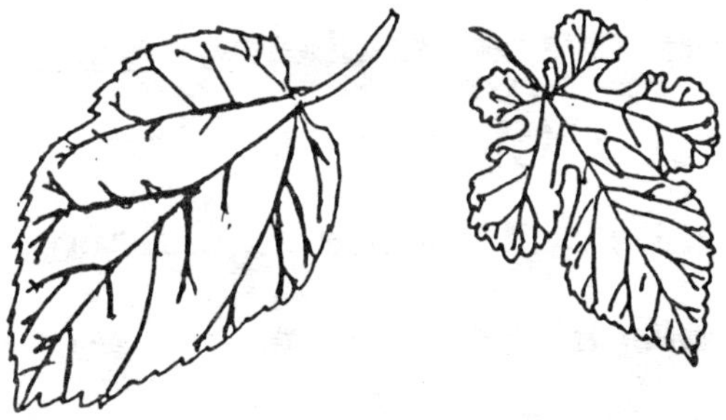

White Mulberry Leaves

White catkins appear on the white mulberry in early spring. Catkins are small greenish flowers in hanging groups.

Some white mulberries have whitish berry-like fruits which appear in summer. Each fruit is about 1/2 inch to one inch long. In some growing areas, this fruit is fed to hogs and chickens. There are fruitless white mulberries too.

BLACK MULBERRY

The black mulberry, Morus nigra, probably originated in Iran. It is sometimes called Persian mulberry. Persia is a former name for Iran.

This mulberry usually grows to about 30 feet high. Catkins are one inch or less.

Black mulberry fruits are about one inch long. They are dark red and when very ripe appear almost black. In some areas the juicy fruit is used to make jam or wine.

Black mulberry leaves can be 8 inches long. They are heart-shaped at the base, hairy on top and downy underneath. In autumn, leaves turn lemon-yellow.

RED MULBERRY

The red mulberry, Morus rubra, is a native American tree. It is common on the East Coast. Look for it in old gardens and along roadsides.

This mulberry has dark red fruits ripening in summer. The fruit resembles

blackberries. Song birds and squirrels are very fond of this fruit.

Red mulberry trees grow from 30 to 60 feet high. Trunk bark is reddish-brown. Lobed leaves are sandpapery above and slightly hairy underneath. Leaves are from 3 to 10 inches long. Sap from trees and leafstalks is milky.

GROW YOUR OWN MULBERRY TREE

Mulberry trees are available in many large nurseries. You can also grow them from cuttings or from seed. In autumn or winter, have an adult make several softwood cuttings (easily cut tree branches), about 6 inches long.

Place each cutting in a container filled with garden soil or purchased potting soil.

Put the container on a sheltered porch or outdoor area that gets sun. Water when soil feels slightly dry. The reason you do several softwood cuttings is to increase your chances that at least one will grow roots. You can move your trees into the ground once they get truly started.

You can also place cuttings directly in the ground, especially if you live in an area which doesn't freeze over in winter.

The trees grow quickly. An adult may have to stake young trees to make certain they don't blow over or snap in a windstorm.

Read a garden book which describes how to stake trees.

Mulberry trees prefer a sunny spot. They do best with regular watering and fertilizing. Large trees can tolerate some drought. Mulberry trees grow along dry beach fronts and in the desert.

Because falling fruit makes a mess, don't put fruiting mulberry trees near a patio or a place where people picnic or children play. The fruit can stain clothing.

Word study: <u>Bombyx mori</u>, the silkworm's Latin name, comes from <u>Bombycidae</u>, the silkworm's moth family. <u>Morus</u> is the Latin name for mulberry.

Chapter Six

OTHER SILKMOTHS PRODUCING SILK

All caterpillars can produce silken threads, although the amount varies. In nature, caterpillar silk threads are used as guidelines to feeding places, as footholds on slippery leaves, and to create cocoons.

However only a few caterpillars produce silk which can be used to make fabric. Besides Bombyx mori, caterpillars most often used to produce fabric silk are: Tusseh silkworms, Chinese oak silkworms, Eri silkworms, and Muga silkworms.

All silk produced by these caterpillars is called "wild silk," because the caterpillars are cultivated in rainforest or jungle rather than controlled conditions. However the use of wild silk is somewhat limited, as the silk from wild silkworms creates a rougher cloth. This is not as easily bleached and dyed as Bombyx mori, mulberry silkworm silk.

TUSSEH SILK

China and India are currently the main producers of Tusseh silk, which has an off-white color. This wild silk generally comes from Tusseh silkworms, offspring of the 5.9 inch wide Tusseh silkmoth, Antheraea mylitta. The Chinese oak

silkworm and the Japanese oak silkworm also make what is called Tusseh silk.

The Tusseh silkworm grows up to 6 inches long. It is pale green on top and darker green below, with a yellow separation line between the two shades. Its primary food is oak leaves, especially white oak.

The greyish-brown cocoon of the Tusseh silkworm is the size of a small hen's egg.

Tusseh silk is naturally tan or brown, making it difficult to dye. It is also less shiney than cultivated mulberry silk. Usually Tusseh silk is blended with other material fibers rather than used alone. It is quite long lasting.

MUGA SILK

The caterpillar of the Muga silkmoth, Antheraea assamensis, (alt: A. assami) produces a golden-yellow silk. This silkmoth is native to India's humid Assam Valley, and at present Muga silk comes only from India. The Muga moth is a light chestnut color. The caterpillar is a medium green with short stiff body hairs. It eats only oak leaves.

ERI SILK

The Eri silkworm, Samia (Philosamia) cynthia ricini, originated in India, but is now found in many parts of the world. It was brought to North America in about

1861, in hopes of getting commercial silk from its threads. Some moths escaped, so the Eri silkmoth is now common, especially in city parks. An alternate name for the Eri silkmoth is Ailanthus silkmoth, because the leaves of this common tree are a favored food plant.

The Eri silkmoth has a wingspan of about 5 inches, but this varies considerably. The moth is brown with white, pink, and tan wing markings.

The Eri caterpillar is green with black dots. It produces a white or reddish silk thread. Fabric made from this thread has a rough texture, and is very long lasting. However the thread is difficult to reel off, or unwind, from the cream-colored, chestnut-speckled cocoon. So Eri

silk is commercially used less than the other wild silks.

SHANTUNG SILK

Shantung silk is produced by the Chinese oak silkworm, Antheraea pernyi. The silk is beige colored. This large tan silkmoth is native to Southern China. Its preferred food is oak leaves. The caterpillar is pale green and lightly covered with yellowish hairs. It has a tan freckled head.

Chapter Seven

SILKWORM FARMING TODAY

Japan was the world's biggest raw silk producer and exporter until the 1970s. Today China produces about 80 percent of the world's raw silk supply. It has more than 10 million silk farmers, and another 1/2 million people employed in making silk fabric.

Although China certainly leads the market, about 34 countries also produce raw silk. The other major producers are Japan, India, Russia, South Korea, and Brazil. France, Italy and Great Britain

produce some raw silk and do silk research. The United States is the leading manufacturer of silk products.

While methods vary in each area, the following is a basic idea of what occurs with mass silk thread production.

Silkworm farmers who regularly raise 1000s of silkworms work under very controlled conditions to get the best quality commercial silk. All silkmoth eggs, often called "seed" or "grains," are inspected under a microscope before being placed in cold storage. In cold storage, the temperature is kept at slightly less than 50 degrees F.

Within the storage site, the air is dry and circulates constantly. This controlled cold storage gives silk farmers the ability to slow down and stagger the hatching period.

In early spring, when mulberry leaves become available, silk farmers put some of the silkmoth eggs in an incubator or climate-controlled room. Within the incubator, the temperature is gradually raised to about 73 degrees F. In about 10 days, silkworms hatch.

After hatching, silkworms are placed on large clean rearing trays. In many places today, rearing trays are fitted with flat bottomless boxes divided into many tiny compartments. One silkworm sets up its cocoon in each compartment.

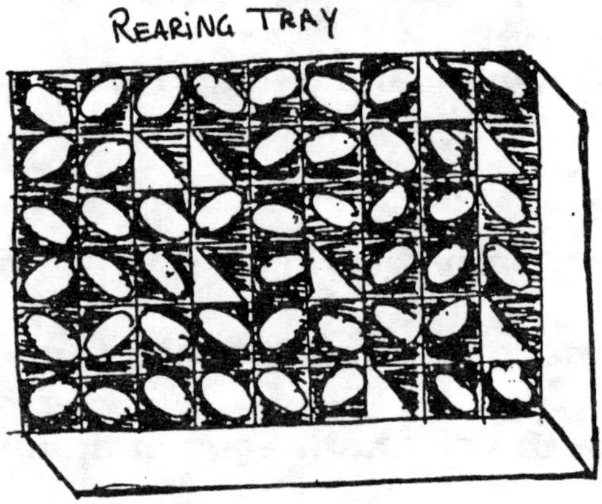

Workers feed chopped mulberry leaves to the silkworms about every two hours. They also make certain the air in silkworm chambers, or rearing rooms, is kept very clean and fresh. Chamber temperature must stay at 75 degrees F. with a 65 percent humidity factor.

Workers are also always on the lookout for silkworm predators, or enemies, as mosquitoes, flies and other insects.

As cocoons are completed, workers harvest or collect them in large straw baskets. Silkworm farmers sell these cocoons. They are paid by cocoon weight and silk quality.

The cocoons are taken to another site or area. Here harvested cocoons are baked or steamed to kill the pupae. Afterwards the cocoons are washed, then soaked in basins of very hot water. This loosens the sticky sericin holding the silk threads together.

Today mechanical brushes beat the cocoons to loosen the mass of short fibres, or “floss” on the outside of the cocoon. This floss is removed.

Now workers locate the end of each cocoon’s continuous silk strand, or filament. Because a thicker thread is easier to handle than very fine individual silk threads, the starting strands from four to seven cocoons are grouped and guided together onto reeling machines. These machines unwind the cocoons.

The resulting thicker silk thread is washed, twisted, coated with a temporary identification dye, and wound into skeins or coiled bundles. This is called "raw silk."

The parts of the cocoon that cannot be reeled are called "waste silk." Two types of silk are made from waste silk: schappe and noil. Schappe or spun silk has a smooth texture. Noil silk has a rougher texture.

Skeins of raw silk are now placed together into a large bundle called a "book." There are about 30 large skeins in each book. The books are packed together to make a "bale." A bale contains about 30 books. Each bale weights about 130 pounds. Transport of silk is done in bale form.

What is made from silk? Sheets, stockings, dresses, kimonos, parachutes, some bike tires, screens, underwear, tapestries, surgical sutures, furniture covering, ties, hats, fly fishing lines, gloves and many other items.

Research continues on ways to improve silkworm productivity. Artificial diets are reducing dependency on mulberry leaves. Japanese agricultural specialists have developed mulberry trees growing in shorter shrub form, making leaf collection easier and faster. Selective mulberry tree breeding gives leaf production for a much longer period.

Lighting in silkworm rearing chambers can now be manipulated to produce an artificial spring/summer throughout silkmuch of the year. This permits, in some silkworm farming areas, eight generations of silkworm per year rather than the normal one generation per year.

DID YOU KNOW?

The raising and keeping of silkworms is called <u>sericulture</u>.

It takes 110 cocoons to make a tie, 630 cocoons to create a blouse, 3000 cocoons to provide silk for a kimono, 800 to make a men's shirt.

It takes 30,000 silkworms to make 12 pounds of silk.

During their growth period, 30,000 silkworms devour about 1350 pounds of mulberry leaves.

Do you remember this nursery rhyme?

"Here we go round the mulberry bush, the mulberry bush, the mulberry bush. Here we go round the mulberry bush, so early in the morning."

Supposedly this began when children would help their parents pick mulberry leaves for silkworms. Why it says "bush" instead of tree, nobody is sure.

Chapter Eight

SILKWORM STUDY PROJECTS

Silkworms are living creatures, and any study projects using them must respect this. Use the greatest of care, and don't cause the silkworms any discomfort. Plan your project thoroughly beforehand. Here are some starter ideas.

1. Does silkworm growth rate change when the amount of light changes?

Raise some silkworms with 2 hours light a day, others with 4, 8, 12, or 16. Do not put artificial light too near the container, as it can overheat the silkworms and possibly bake them. You can just cover the containers for varying

periods to change lighting (all covers must have ample airholes) or use a light timer.

2. Does silkworm growth rate change with different light colors?

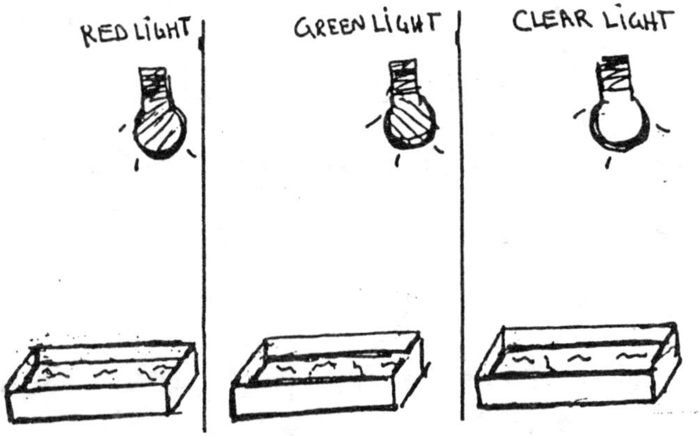

Keep silkworms in three different containers: one with a red light, one with a green or blue light, and one with a clear light.

3. Does silkworm growth rate change with temperature variations?

Use several containers. Shoe boxes are fine. Put 4 to 6 silkworms in each. The number should be the same in each container. Put one container in a cool (not cold) place, another on an average temperature shelf, and another in a warmer (not hot) room section. Feed each grouping the same number of leaves at the same intervals.

Which box has the biggest silkworms at the end of each week? In which box do the silkworms begin their cocoon first? Last?

4. How far will a silkworm crawl to reach food?

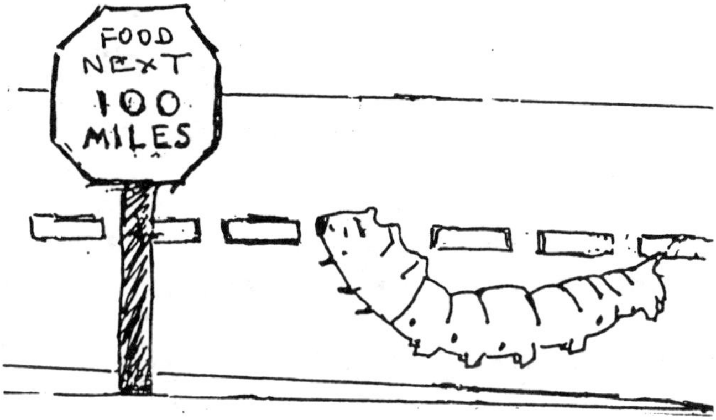

Place a few silkworms in a separate container. Be careful not to disturb any silkworms that are molting. Put food leaves about 2 inches away. When these have been eaten, place food leaves 4 inches away. Keep doing this. At what point will the caterpillars stop trying to reach the leaves?

Some reference texts say maximum searching distance is 12 inches. Others say it's only a few inches. What do you think? Does it depend on the silkworm size or age? That's another project.

5. Which type of mulberry leaves make the best silkworm food?

You need at least two different types of mulberry leaves: red, white, or black, to do this project. Some studies show that silkworms feeding on black mulberry leaves create lighter cocoons with less silk, develop more slowly, and don't live as long. With black mulberry feeding, did you find this to be true? What about a comparison between the other leaf types?

There are nine commercial grades of cocoons. Can you find these among your cocoons?

1. **Good cocoons**
2. **Pointed cocoons (not good for thread reeling)**
3. **Cocalons: larger than normal**
4. **Duppions: doubles, where two caterpillars are in one cocoon, usually a male and a female.**
5. **Soufflon: silk threads are loosely meshed, cocoon may appear transparent.**
6. **Perforated: cocoons are broken in some way.**
7. **Good choquettes (dead pupa inside, cocoon does not rattle.)**
8. **Bad choquettes (rotten).**
9. **Calcinated cocoons: contain petrified or hardened pupae that rattle inside the cocoon.**

7. Cocoons containing female pupae weigh more. Can you divide your cocoons into male and female according to weight?

Put the cocoons you guess as male or female in separate containers so you can track the results. You will need some way to weigh the cocoons. Can you create a baancing scale, or obtain a sensitive weighing scale?

Many silkworm experts can just hold the cocoon and tell from its weight whether it's male or female. You can try this.

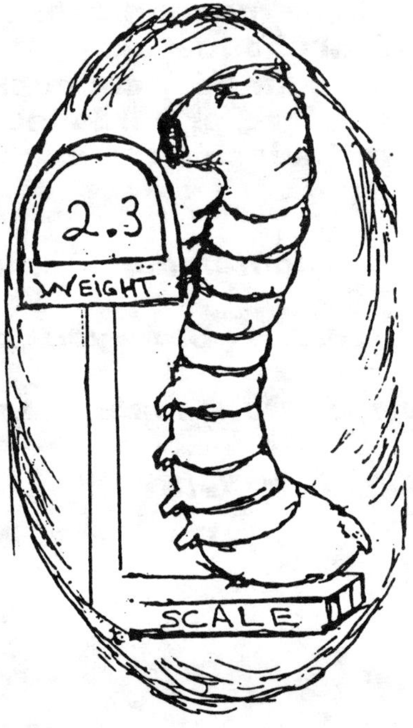

Put the cocoons in separate male-female containers so you can track the results. An alternative to separate containers is finding a way to mark each cocoon with your guess as to whether it is male or female. How accurate are you?

8. Do the silkmoths coming out of different colored cocoons look the same?

You need to place gold, yellow, or other color cocoons separately. That way you can observe the silkmoth emerging from each different colored cocoon.

9. All texts say silkmoths cannot fly. Test this yourself. If they can't fly, how far can they hop? Travel across a room? Do males or females seem to travel furthest?

10. Do silkmoths emerging from varying color cocoons produce eggs which will give that same color cocoon the next season?

This is a continuing project in genetics and can take one or more years to document. You will have to control the mating procedure to get accurate results.

***11. Studying pupal growth*: Using a razor blade, an adult can very carefully cut open the ends of a few cocoons and very gently remove the pupae within. Females will be larger than males. Are there any other differences?**

Looking carefully you may see the bottom of some pupae are notched or irregular. These pupae are female.

Now very carefully place the cocoon halves back together. If done correctly, the pupae will continue normal growth and the silkmoths will emerge on schedule.

12. Unwinding silk from your cocoons.

You cannot unwind the silk all in a single thread if the moth has emerged from the cocoon. The hole made by the emerging moth cuts the silk thread in many places. That's why, for business reasons, the pupae inside the cocoon are destroyed before the silkmoth exits. This leaves the mile-long thread all in one piece.

If you don't want to destroy the pupa, there is a possible option. If you have many cocoons, you may find that some remain intact. For some reason, the moth does not emerge. So you can use the intact cocoons for silk unwinding.

Another method of silk unwinding from the cocoon:

Pick off the cocoon's outer fuzz. Then soak the cocoon in warm water for 30 minutes. The silk threads will loosen. With a dissecting needle or long sewing needle, find a single thread that pulls easily.

With the cocoon still in the water, wind this silk thread onto a pencil or slim round stick. You will get some silk thread, although not a lot. Now remove the cocoons from the water. Replace them in the hatching container. Silkmoths may come out.

ENVIRONMENTAL CAUTION

Scientists believe there are no more wild silkworms. Why? Early silkworm farmers kept only the breeding moths that had the least tendency to fly, hide or wander away. This made the farmer's job easier.

From time to time, some of the captive moths escaped and mated with their wild relatives. The result was a very sluggish wild moth. In the wild, there are many moth enemies, including birds. The sluggish, nonflying, nonhiding moths were quickly eaten. Over time, no silkworm moths remained outdoors except in protected silk farms.

You must protect your silkmoths too. Never release silkmoths or silkworms into the wild. They will almost certainly be eaten by birds or other predators.

GLOSSARY

Advocate pleads in favor of a cause
Ambush to wait and attack by surprise
Antennae insect sensory organs or "feelers"
Artificial not real, fake
Attendant a regular follower
Capsule a small case
Commercial business
Cocoon silky envelope spun by many insects
Cultivate to promote growth
Continuous uninterrupted
Deciduous loses leaves in winter
Emerge come out
Extinct no longer exists
Fertilize joining of sperm and ovum
Fleece coat of wool covering a sheep or similar animal
Frass waste material
Ganglia nerve cell groupings
Gland cell group producing a specific product
Grove a cluster of trees
Grain another name for silkworm eggs
Gut intestine
Humid moist
Incubator climate-controlled container
Industry manufacturing business
Instar an insect stage between molts
Metamorphosis a marked change of form
Molt to shed, as skin
Larva caterpillar
Legend a story based on tradition

Malpighian tubes an insect digestive mechanism

Manipulated to skillfully manage or change

Ovary female reproductive gland

Penalty punishment

Pheromone special chemical odor
Predator enemy
Pupa insect form between larva and adult stages
Seed another name for silkworm eggs
Sericin sticky fluid joining silk strands
Sericulture rearing silkworms to make silk
Skein coiled bundle of yarn or thread
Sperm male reproductive cells
Spinneret tube under silkworm's mouth
Technique method of doing
Technology industrial knowledge
Thrive to succeed, to increase

BIBLIOGRAPHY

Ahmad, Humma. China's Silken Success, World Press Review, December 1987, pp. 49.

Carolina Biological Supply Company, Culturing Silkworms, Oregon, 1986.

Clausen, Lucy. Insect Fact and Folklore, Macmillan, London, 1967.

Collins, Michael, et Weast, Robert. Wild Silk Moths of the United States, Collins Radio Company, Iowa, 1961.

Crotch, W.J.B. A Silkmoth Rearer's Handbook, Dept. of Agriculture, The University, Leeds, Yorks, 1956.

Currie, Mr., trans. Silk (La Soie), International Silk Association, Lyon, France, 1994.

Feltwell, John. The Story of Silk, St. Martin's Press, N.Y., 1990.

Holland, W.J. The Moth Book, Dover Publications, N.Y., 1956.

Hyde, Nina. Silk--Queen of Textiles, National Geographic, January 1984. pp 5 - 48.

Klots, Alexander. 1001 Questions Answered About Insects, Dover Publications, NY, 1977.

Lutz, Frank E. Field Book of Insects, G.P. Putnam's Sons, New York, 1948.

Petrides, George A. A Field Guide to Trees and Shrubs, Houghton Mifflin, Boston, 1958.

Ritchie, Carson. Insects, the Creeping Conquerors, Elsevier/Nelson Books, NY, 1979.

Silk, The Silk Association of Great Britain, 1994.

Sunset Western Garden Book, Lane Publishing, Menlo Park, CA, 1988.

Unite Nationale Sericicole bulletin, Ministry of Agriculture, France, 1994.

Villiard, Paul. Moths and How to Rear Them, Funk & Wagnalls, N.Y., 1969.

PLACES TO WRITE or visit

WORLD OF SILK
David Evans and Company
Bourne Road, Crayford, Kent
England. This is silk screen printing center and factory shop.

THE SILK MUSEUM
The Heritage Centre
Roe Street, Macclesfield
SK 11 6UT Great Britain
museum, movie, tours
mill historical exhibition

WORLDWIDE BUTTERFLIES
& LULLINGSTONE SILK FARM
Compton House near Sherborne
Dorset, DT9 4QN, Great Britain

The silk farm supplying silk for the Queen's wedding dress and other royal ceremonies. Silkworms in action, film, factory process.

WHITCHURCH SILK MILL
Whitchurch, Great Britain
working silk mill, historic looms, water wheel.

THE WORKING SILK MUSEUM
New Mills, South Street
Braintree, Essex CM7 6GB
Great Britain

INDEX

ABOUT THE AUTHOR

L. Patricia Kite holds a biology teaching credential, a B.S. from U.C. Medical Center, and a master's degree in journalism. She is a member of NSTA and SCBWI.

Prior book honors include a National Parenting Press (NAPPA) gold award, inclusion in the 1997 YALSA quick picks for Reluctant Adult Readers list, and inclusion in Science Books & Films Best Books for Children 1992-1995 recommended reading list.

Kite specializes in writing about insects, plants, and plant-insect interaction, for both young children and adults. She has raised silkworms for many years, as well as judged many science fair competitions.

For information on school, library or book store presentations, write to L. Patricia Kite, 5318 Stirling, Newark, CA 94560.

WARNING-DISCLAIMER

This book is designed to provide information in regard to Silkworms. Every effort has been made to make text as accurate as possible within book length parameters. However there may inadvertly be mistakes both typographical and in content. Therefore book should be used only as a general educational guide. The author and LPK Science shall have neither liability nor responsibility to any person or entity in regards to any loss or damage caused, or alleged to be caused, directly or indirectly by the information contained in this book.

If you do not wish to be bound by the above, you may return this book to the publisher for a full refund.

To order additional copies of SILKWORMS, complete the information below.

Ship to: (Please print)

Name_______________________________________

Address_____________________________________

____

City, State,Zip________________________________

Day phone __________________________________

____copies of SILKWORMS @$8.95 each **$ _____**

Priority mail postage (1-2 books) **$3.00**

CA residents add 74 cents tax **$______**

Total amount enclosed **$______**

Make checks payable to LPK Science
5318 Stirling. Newark, CA 94560